STEAM & Me™
SHARKS

L. J. TRACOSAS

Starry Forest Books

SCIENCE · TECHNOLOGY · ENGINEERING · ARTS · MATHEMATICS

Draw a super-smart robot. Create your own wind energy. Find out if your teeth are as sharp as a shark's. Go back in time to the world of dinosaurs or rocket into space. Power up that scientific brain of yours with STEAM & Me!

Photos, facts, and fun hands-on activities fill every book. Explore and expand your world with science, technology, engineering, arts, and math.

STEAM&Me is all about you!

Great photos to help you get the picture

New ideas sure to change how you see your world

Flat Family
Sharks are related to rays and skates, flat fish whose skeletons are also made of cartilage.

A shark's skin is dotted with pointed scales called *dermal denticles*. Stroke a shark from head to tail and it will feel smooth. But stroke it in the opposite direction and it will feel like you're petting sandpaper. Scratchy!

What is a shark?

A shark is a type of fish—but it's different from your goldfish. Like all fish, sharks live in water. They breathe by passing water through their gills. Other fish have just one gill slit on each side of the head. Sharks have five or more. Most fish skeletons are made of bone. But a shark's skeleton is made of a more flexible material called *cartilage*.

Say "Cheese"!
Sharks aren't the only fish that have teeth. The pacu fish's teeth look like yours. Smile!

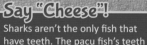
STEAM&Me Your skeleton is made of bone. But the human body also has cartilage in it. Lightly touch the end of your nose. Now gently wiggle and bend the upper tips of your ears. That's cartilage, the same flexible stuff that a shark's skeleton is made of.

Fascinating facts to fill and thrill your brain

Hands-on activities to spark your imagination

Dive in!

What do you think of when you hear the word *shark*? A fin moving through the waves? Or big, chomping jaws and sharp teeth? Once you dive into the amazing world of sharks, you'll find there's a lot more to them than their pearly whites. There are more than 400 **species**, or types, of shark swimming in the world's oceans. And new types are discovered all the time.

In boxes like these throughout the book, learn new ways to think about a shark's life . . . and yours!

STEAM & Me

Tiger sharks are one shark species.

Way Old!

Some shark fossils are more than 400 million years old. That means sharks swam in the oceans millions of years *before* dinosaurs even existed!

Flat Family

Sharks are related to rays and skates, flat fish whose skeletons are also made of cartilage.

Say "Cheese"!

Sharks aren't the only fish that have teeth. The pacu fish's teeth look like yours. Smile!

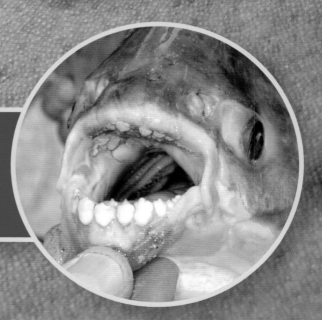

A shark's skin is dotted with pointed scales called *dermal denticles*. Stroke a shark from head to tail and it will feel smooth. But stroke it in the opposite direction and it will feel like you're petting sandpaper. Scratchy!

What is a shark?

A shark is a type of fish—but it's different from your goldfish. Like all fish, sharks live in water. They breathe by passing water through their gills. Other fish have just one gill slit on each side of the head. Sharks have five or more. Most fish skeletons are made of bone. But a shark's skeleton is made of a more flexible material called *cartilage*.

STEAM

Your skeleton is made of bone. But the human body also has cartilage in it. Lightly touch the end of your nose. Now gently wiggle and bend the upper tips of your ears. That's cartilage, the same flexible stuff that a shark's skeleton is made of.

Where do sharks live?

Sharks live in waters all around the world. Their **habitats** differ as much as people's habitats do. Some sharks live in the open ocean, miles away from shore. Others live in shallow water. Some live along coral and rock reefs, where there are plenty of fish and sea creatures to eat.

Reef sharks find plenty to eat in the coral reefs that are their home, or habitat.

Would you rather be a nurse shark that stays close to home, or a great white shark that travels more than 2,500 miles every year?

STEAM & Me.

So Chilly

A Greenland shark lives in Arctic waters, which is a very cold habitat.

It's Dark Down Here!

Deep-sea sharks like the goblin shark live thousands of feet under the water's surface, where it's pitch black.

The whale shark can be 40 feet long, or as long as a school bus. It is one of the most gentle sharks.

Big Shark
Basking shark: about 29 feet long

Small Shark
Pygmy shark: about 8 inches long

The
bull shark is
about 7 feet long.
It is the most likely
to attack humans.

Sharks can be really, really BIG. Sharks can be quite small.

The biggest shark in the ocean is the whale shark. It's not a whale at all, but it's as big as one. You eat about 3 pounds of food a day. A young whale shark eats about 46 pounds of **plankton**, tiny animals and plants that float in the water, every day. Sharks can also be very small. A pygmy lantern shark is about as long as a pencil.

Sharks can be SUPER swimmers.

Sharks are built for swimming. Because of its strong muscles, the mako shark powers through the water in bursts of more than 50 miles per hour. That's as fast as some cars go on the highway. *Zoom!* The giant whale shark moves more slowly, but it swims long distances. One whale shark was tracked through the Pacific Ocean, traveling more than 12,000 miles. That's like going all the way across the United States and back again—twice!

The mako shark has a streamlined shape, so water flows around it easily.

The mako shark steers with its top, or *dorsal*, fin and its side, or *pectoral*, fins.

Powerful muscles move the mako's tail, or *caudal* fin, through the water.

Put your hand in a tub of water with your thumb facing up. Move it through the water. Do you feel the water pushing back? Now move your hand through the water with your palm facing down. It's easier to move, right? That's streamlined!

STEAM *& Me*

Shaped for Speed

Trains and airplanes are also streamlined to go fast.

Fish Out of Water

Epaulette sharks do more than swim to get around. They sometimes use their fins to "crawl" over reefs.

Super Tails

Thresher sharks' super long tails are super streamlined!

See eye to eye with a shark.

Sharks' eyes are on the sides of their heads. They can see almost all the way around themselves—above, below, both sides, and some of what's behind them. Some sharks, like an angel shark, have small eyes, with a slit-shaped pupil. The blue shark's big eyes look almost black. They help it see well in the bright water near the ocean's surface. Deep-water sharks like the megamouth also have big eyes that help them see in the ocean depths where it is pitch dark.

Seeing in the Dark

A shark's eye is similar to a cat's eye. It has special tissue that helps it see in the dark.

Blink!

Sharks close their eyes when they bite so food doesn't get in their eyes. Some have a special membrane that acts like an extra eyelid for protection.

The blue shark's eyes look solid black, but when you are close, you can see their pupils.

If a tooth falls out of a shark's mouth, another one is right behind to take its place. A great white shark can go through 20,000 teeth in a lifetime!

Mmm . . . Crunchy!

Nurse sharks have flat teeth for grinding. They can eat animals with shells, like crabs and mussels.

Don't Worry

Most sharks don't bite humans. Shark attacks are rare.

Sharks are predators.

A great white shark is one of the biggest **predators**, or hunters, in the world. It has big teeth for catching big **prey**, like seals. Sharks that eat only other animals are *carnivores*, or meat-eaters. There's a different kind of hungry shark to eat just about every type of sea creature, from the tiniest plankton to the slimiest octopus.

Open WIDE

A basking shark is a filter feeder. It opens its big jaws, gulps water, and then pushes the water out, filtering it through its gills so that plankton remains in its mouth. Then it swallows the plankton—yum! Its teeth are tiny, because the shark doesn't need them to chew.

STEAM & Me

What kind of teeth do you have? With a clean hand, feel your teeth. Do you have any sharp ones, like a great white's? Do you have any flat ones, like a nurse shark's? Why do you think your teeth are shaped the way they are? How many teeth do you have?

No noses needed.

Most sharks move all the time to keep water flowing through their gills. That's how they breathe! Many sharks have five gills. Some, like the broadnose sevengill shark, have up to seven.

Another Way to Breathe

Some sharks, like horn sharks, can suck water in through their mouths and over their gills. That way, they can stay still on the bottom of the seafloor.

Count the gills. How many does this shark have? (Hint: It's called a sixgill shark.)

Sniff!

Sharks have a very keen sense of smell. It's so good that some sharks can smell one drop of fish oil in an Olympic-size swimming pool.

Can you point out the nostrils on this lemon shark?

So what does a shark use to smell?

A shark doesn't have a nose like you do. But sharks do have nostrils. You can see them on the bottom side of their snout. Because sharks breathe through their gills, their nostrils are just for smelling. A tiger shark can sniff out a sea turtle from as far as a football field away.

22

Listen up!

A shark has two ears. But they're hard to see. The ears are tiny holes behind each eye. Sound enters through the ear holes. Then sound waves make little hairs and fluid in the ears vibrate. Sharks have very good hearing. They listen for the sounds of prey, like the splashing of a seal in the water.

Feel the Vibrations

Hearing and smell help sharks find prey. Sharks can also sense prey with special *pores*. These little holes in their skin alert them to other animals' movements.

How does a shark hide?

Sharks are sneaky hunters. They use **camouflage**, special patterns on their skin, to blend in with their habitat. Then they surprise their prey. The wobbegong shark is one of the ocean's best hiders. It's also called a carpet shark, because it lies flat on the seafloor like a rug. It even has tassels around its mouth. To other sea creatures, those tassels look like seaweed.

Look Up

Sharks like the mako that hunt in the open ocean have lighter skin on their bellies. That makes them blend in with the sunlight so prey below have a hard time seeing them.

People sometimes wear camouflage clothes to better blend in with the background. What color clothes would help you blend in with your backyard? What about on a sandy beach?

STEAM & Me

Look Down

Sharks often have darker skin on top. It blends in with the dark water below them. Prey up above have a harder time spotting them.

Where's the wobbegong? Wobbegong sharks use camouflage to surprise crabs, octopuses, and fish. First they lie very still. Then they snap and catch their prey.

Speak a shark's language.

A shark can't speak. But it can use its movements and other body language to communicate. That helps other sharks—and people—know what it might be thinking. A shark swimming steadily and making slow turns is probably at ease. But if it's arching its back, lowering its pectoral fins, and making fast turns? Those signs mean a shark feels threatened. Or it's ready to hunt.

This shark's arched back is a way of saying, "Stay back! You're making me nervous!"

Schools of Sharks

Some sharks, like great hammerhead and reef sharks, swim in groups called *schools*. Schooling sharks tend to use more body language than sharks that swim alone.

Stay Away!

The swell shark can suck water into its body and puff up to make it look bigger. It's telling predators to stay away.

Talking helps you communicate, or share what you're thinking and feeling. Body language does that, too. Think about what happens when you're worried. What does your face do? Do your shoulders feel different? That's your body language! Now think about feeling happy. What does your face do then? How else does your body language change? Are there other ways that people communicate without talking?

Wake Up!

Many sharks are *nocturnal*. That means they hunt at night. Often, sharks rest during the day and are very active after dark.

These nurse sharks are resting on the seafloor.

Good night, shark.

At the end of the day, you're tired and ready to sleep. Not so for a shark. Some sharks must always be moving to keep water flowing over their gills. Instead of going to sleep, a lemon shark takes a rest. During this time, it's awake and moving, but slowly.

Flip

Researchers sometimes turn a shark upside down to study it. This makes the shark less active, almost like it's sleeping. Scientists aren't sure why this happens.

So many sharks!

Sharks are amazing creatures with really cool features. You've learned how they're like you, and how they're not. You've read about 25 sharks in this book. Which is your favorite?

Look at the pictures on this page. Can you go back and find these sharks in this book? What is each shark's name? Can you say one fact you learned about each shark?

31

Glossary

Learn these key words and make them your own!

camouflage: a design or coloring that makes something harder to see. *The wobbegong shark uses camouflage to blend in with the seafloor.*

habitat: a place that an animal or plant lives and grows. *A reef shark's habitat is a coral reef.*

plankton: tiny animals and plants that live in the water. *Big whale sharks eat lots of little plankton.*

predator: an animal that hunts and eats other animals. *The great white shark is an excellent predator.*

prey: an animal that is eaten by another animal. *Crabs are one of the nurse shark's favorite prey.*

species: a group of animals that are similar to one another and can produce young. *A blue shark is one species of shark, and a swell shark is another species of shark.*

For Miles, who is helping to save the whale sharks.

ASP: Alamy Stock Photo; IS: iStock; SS: Shutterstock. Cover, Sergey Uryadnikov/SS; 4-5, Matt9122/SS; 5, Steve Bower/SS; 6, (UP) Hoiseung Jung/SS; 6, (LO) Pedro Luz Cunha/ASP; 6-7, stockpix4u/SS; 8-9, frantisekhojdysz/SS; 9, (CTR) Nature Picture Library/ASP; 9, (LO) Paulo Oliveira/ASP; 10, (LO LE) Martin Prochazkacz/SS; 10 (LO RT) Nature Picture Library/ASP; 10-11, Krzysztof Odziomek/SS; 12-13, Howard Chen/IS; 13, (UP) scanrail/IS; 13, (CTR) jcapaldi/Flickr; 13, (LO) bearacreative/SS; 14, OlenaKlymenok/IS; 15, (UP) Greg Amptman/SS; 15, Howard Chen/IS; 16, Sergey Uryadnikov/SS; 16, (CTR) Yann hubert/SS; 16, (LO) SylwiaDomaradzka/IS; 17, Martin Prochazkacz/SS; 18, Kirk Wester/IS; 18-19, Greg Amptman/SS; 20, Simon Biffen Photography; 20-21, Mok Wai Hoe/SS; 22, art nick/SS; 23, Havoc/SS; 24, Martin Prochazkacz/SS; 25, (UP) A Cotton Photo/SS; 25, Longjourneys/SS; 26, Tomas Kotouc/SS; 26-27, Constantinos Petrinos/NaturePL.com; 27, Mark Conlin/VWPics/ASP; 28, nicolasvoisin44/SS; 28-29, skeeze/Pixabay; 29, Cultura Creative/ASP; 30, (UP 1) Sergey Uryadnikov/SS; 30, (UP 2) Krzysztof Odziomek/SS; 30, (LO 1) Longjourneys/SS; 30, (LO 2) Natalie Ruffing/IS; 31, (UP LE 1) Nature Picture Library/ASP; 31, (UP LE 2) bearacreative/SS; 31, (LO LE 1) Martin Prochazkacz/SS; 31, (LO LE 2) Howard Chen/IS; 31, (UP RT 1) Dusseau Photo/SS; 31, (UP RT 2) Kirk Wester/IS; 31, (LO RT 1) Matt9122/SS; 31, (LO RT 2) Paulo Oliveira/ASP; 32, (CTR) Andrea Izzotti/SS; 32, (LO) wildestanimal/SS; Back cover, (UP) Havoc/SS, (LO LE) wildestanimal/SS, (LO CTR) Andrea Izzotti/Shutterstock